House On The Hill

House On The Hill

A Metaphysical Journey Into The Glass Bead Game

Mark Megna

To order additional copies of this book, contact:
Xlibris Corporation
1-888-795-4274
www.Xlibris.com
Orders@Xlibris.com
121626

Foreword

Purpose of Study into The Glass Bead Game

The field of quantum theory opened up new ideas of where science is going. With new understanding of what our universe truly is, we have begun to look at our existence in a completely new way! This "new" thinking is what has inspired me to research the consciousness of humans, with hopes to gather a better understanding of how exactly we fit into this thing we call reality.

For thousands of years, priests, sages, monks, and philosophers have all referenced cases of this higher state of consciousness. The Christians call it uniting with God. Eastern thinkers may call it the "awakened" mind. Philosophers like Friedrich Nietzsche referred to this state as "Overman". So what is this consciousness that these people speak of? My intent with studying philosophy is to bring the concepts of these great thinkers of the East and West in order to create a new understanding of what these people have written and talked about throughout the existence of man. Therefore, with this knowledge being so easily accessible to people in this information era, I feel that the human species is approaching another stage of evolution. This next step in our development will not be a physical adaptation; but rather a conscious evolution into a higher conscious state of being. I intend to study this new territory of our existence and plan to uncover a comprehensive way for us all to understand and obtain Cosmic Consciousness.

Tony Megna CSCS

Contents

Foreword..5

Hermann Hesse—The Glass Bead Game...........................13

Ayn Rand—Objectivism—Self Interest—Realism..................14

Alan Watts—Eastern Mysticism...................................15

Plato—Forms—Idealism..16

Parmenides—One...17

Zeno—Paradoxes...18

Douglas Hofstatder—Godel, Escher, Bach.......................22

Aristotle—Physics—Realism—Metaphysics......................23

Anaxagoras—Everything is in Everything........................24

Bertrand Russell—Continuity....................................25

Kurt Gödel—Incompleteness Theorem............................26

Werner Heisenberg—The Uncertainty Principle...................28

Erwin Schrödinger—Schrödinger's Cat—Superposition...........29

Georg Cantor—Infinity...30

F. H. Bradley—Appearance and Reality..........................31

George Berkeley—To Be is To Be Perceived......................32

Gottfried Leibniz—Monads.......................................33

Euclid— "A Point Is That Which Has No Extension" 34

Democritus—Atomism and Void 35

Albert Einstein—Relativity 36

John Paul Sartre—Being and Nothingness 37

Gary Zukav—The Dancing Wu-Li Masters 38

John Searle—Philosophy of Mind 39

Brian Green— String Theory and The Elegant Universe 40

Immanuel Kant— Critique of Pure Reason 41

David Hume—Metaphysics 43

Max Planck—Planck Units—Discreteness 44

Henri Bergson—Intuition 45

Charles Siefe—Zero—A Biography of a Dangerous Idea 46

Paul Davies—About Time 47

Joseph Mazur—The Motion Paradox 48

Rene Descartes— "I think, therefore, I am" 49

Fritjof Capra—The Tao of Physics 50

Julian Barbour—The End of Time 51

Stephen Hawking—A Brief History of Time 52

John Mctaggart—The Unreality of Time 53

Roger Penrose—The Emperors New Mind 54

Alfred North Whitehead—Process and Reality 55

Richard Bach—Jonathon Livingston Seagull—Limits 56

Isaac Newton—Laws of Motion 57

Michelangelo—The Pieta, The Sistine Chapel, and The Statue of David 58

John Keats—Beauty is Truth, Truth Beauty 59

Lisa Randall—Warped Passages—Hidden Dimensions 60

Thomas Nagel—A View from Nowhere... 61

Thomas Young—Double Slit Experiment... 62

John Stuart Bell—Incompleteness Theorem—Entanglement............................... 63

Louis DE Broglie—Wave-Particle Duality of Matter 64

John Archibald Wheeler—Law without Law and Black Holes 65

Neils Bohr—Theory of the Atom 67

Bryce DeWitt—Many Worlds Theory 68

Hugh Everett—Many Worlds Theory 69

Leo Buscaglia—Love.. 70

Erich Fromm—The Art of Loving 71

Heraclitus—Flux—Eternal Change 72

Georg Hegel—Idealism 73

Fredrick Nietzsche—Nihilism—Belief in nothing absolute— "Overman" 74

Jacob Bernoulli—Infinity—A finite set contains an infinite amount of members 75

M. C. Escher—Escher Tiling's—An enclosed image that contains an
    infinite amount of images 76

William Blake—Eternity in an hour 77

Blaise Pascal—Man emerges from nothingness... 78

Martin Heidegger—Being and Time 79

Aldous Huxley—A Brave New World.. 80

Mahatma Gandhi—Non-Violence 81

Jesus—Salvation.. 82

Buddha—Zen and Nothingness... 83

David Hilbert—Space... 84

Johannes Sebastian Bach—Symphony of the Universe 85

Ludwig van Beethoven—Music Without Sound 86

Wolfgang Amadeus Mozart—Perfection... 87

Jacqueline Du Pre—Cellist... 88

Andre Gregory—My Dinner with Andre.. 89

Stevie Ray Vaughn—Musician 90

Glenn Beck—The Gandhi of Our Time 91

Neil Young—Cortez The Killer... 92

Patti Smith—Rock and Roll Nigger and Poet 93

Orson Wells—Fabian Society 94

Michio Kaku—Buddhism and Quantum Mechanics.. 95

George Orwell—1984 96

Dali—Surrealism 97

Leonardo Da Vinci—Magus and Grand Master of a Secret Esoteric Society 98

Norman Vincent Peale—The power of positive thinking.................................. 99

Conclusion 100

References.. 101

Dedicated to Tony Megna who understands and appreciates the profound significance of The Glass Bead Game.

This book was inspired by the book "The Glass Bead Game" by Hermann Hesse. This is an "entelechy" of a theoretically possible "coming to be" and actualization of an ultimate existence. This is the synthesis of all of the great minds and disciplines in the history of philosophy, religion, science, mathematics, music, art, literature, poetry, painting, sculpting, and theatre, etc. to name a few to create an alternative supreme reality. It is an intellectual and artistic "Hall of Fame" of sorts. The study and realization of the self at a higher state of consciousness came to be in writing this book. A creative freedom was exercised without concern for convention, criticism, or the mundaneness of everyday pragmatic ways of thinking in society and limited viewing of the world. In one respect you can say that this is a history of metaphysics but it is also the synthesis of all of the profound ideas that fuse into one ultimate conception of reality. This glorified reality and act of creating it is at the foundation of my Glass Bead Game in my metaphorical "House on the Hill."

The following is a compilation of all of the great thinkers that I have had the luxury and privilege to read and incorporate into my sanctuary. Much of the book is comprised of the thoughts and ideas of these great minds which magnify the artistic glory of The Glass Bead Game. I, certainly, could not say it any better than they have said it themselves but have tried to elaborate on its significance to my Glass Bead Game. Although many of the reproductions of these great men and woman's work may seem as contradictory to each other they are all significant in their own way towards the ultimate concept of Truth. Moreover, I feel that I have almost attained immortality myself in writing this book since it has been done in that timeless type of fashion and the "the essence of it will forever be." In a way this manuscript is considered my living Bible or I Ching in which I am able to refer back to it and build upon further creations of the reality of The Glass Bead Game. It is an open system.

Hermann Hesse—The Glass Bead Game

Hermann Hesse believed in the eastern way of thinking and mysticism in which he viewed reality from a different perspective than the western ways of thinking. It pursues a state of nirvana through the study of the self and reality by shedding the ego and becoming one with the experience of reality. Similar to the ecstasy that a mountain climber experiences when he climbs Mount Everest in which he becomes one with the event. Time and the self disappear and merges through a focus on the task at hand. In the book, the setting of The Glass Bead Game is a secret society called Castalia where Joseph Knect was a able to study mathematics and music and pursue a higher state of consciousness. This was expanded to include all the disciplines in which Joseph Knect was able to synthesize the history of all knowledge to create an ultimate reality. As it turns out the game is in actuality real and is the study of himself and his journey towards a perfect existence. Many of his other novels such as Journey to the East, Steppenwolf, Narcissus and Goldman, and Damien were Hermann Hesse disguised as fictional characters in which he was able to live a life of fantasy in which he could seek immortality and self-discovery.

"These Games were little dramas, in structure almost pure monologues reflecting the imperiled but brilliant life of the author's mind like a perfect self-portrait."

Hermann Hesse—The Glass Bead Game

Ayn Rand—Objectivism—Self Interest—Realism

Ayn Rand developed the objectivist philosophical system in which the pursuit of one's own happiness is the highest goal of man through the pursuit of maximum human self-achievement. She believed in glorification of the ego which was the opposite of the eastern way of thinking in which believed in the shedding of the ego as the ultimate state of existence. She believed that this losing of the self was evil and self-destructive. Coerced altruism was especially despised by Rand since it limited man's freedom to pursue one's own goals. Unfettered freedom was one of her top guiding principles that she built the objectivist philosophy on. This included economic, intellectual, and artistic freedom. No state had any moral or legal right to hinder or coerce you in any way. This included all forms of coerced taxation in which she considered it to be legalized stealing. In her novel Atlas Shrugged she had all of the industrialists and productive members of society disappear to form their own secret society free from the immorality and evils of liberal progressive socialists. There they were free to pursue the perfection of man one individual at a time through a system of self-reliance.

It was Ayn Rand's development of the Objectivist Philosophy which emphasizes self-interest, self—achievement, self—reliance, and most importantly the freedom to glorify one's own ego that makes her a necessary member of The Glass Bead Game but also as one of the greatest woman to ever live!

Alan Watts—Eastern Mysticism

Alan Watts was an expert on all eastern and western religions and attempted to compare and contrast the various views. He believed that the west was dogmatic whereas the eastern forms of mysticism allowed for the pursuit and study of reality and the self as a path to a state of Nirvana. Alan Watts talks about a "lightning strike" in which you know instantly with 100% certainty the true nature of reality existing within a yin and yang duality of oneness.

Alan Watts spent his entire life seeking a higher state of consciousness which he explained in artistic detail in his countless books and lectures. He created his own unkept secret haven of intellectualism outside of the mainstream of society. His ability to compare and contrast eastern and western ways of thinking was similar to comparisons of the wave-particle duality to reality and existence.

"You don't look out there for God, something in the sky, you look in you."

Alan Watts

Plato—Forms—Idealism

Plato believed that our limited minds were living in a virtual illusion in which we could not perceive the perfection of ultimate reality. He believed that his platonic forms were the ideal that existed outside the confines of space and time. Plato developed "The Academy" which was a form of The Glass Bead Game and foundation of our current university system where higher pursuits of learning could take place in setting outside the demands of everyday society. He expanded his desire for a perfect society in his fictitious creation of Atlantis.

Plato had great historical significance on the idealistic teachings of future philosophers such as Kant, Hegel, Berkeley, and so forth which ultimately recognized the human mind as having a special significance in the universe. An importance so high that some consider it to be the foundation and creator of all reality.

Parmenides—One

Parmenides believed in one immutable, eternal, indivisible reality which transcended the limited world in which we perceive. He stated that "being is so elusive that it even transcended itself."

"The last line of Plato's Parmenides "Then one cannot be anywhere, either in itself or in another" has special meaning in the Buddhist ways of thinking in which man seeks a definition and meaning of himself but has yet to find it.

To a large extent I considered "Metaphysics of Being" a modern interpretation of Parmenides. It merged his thinking with many idealist philosophers, quantum mechanics and Buddhism. The idealists such as Berkeley believed that everything was in the mind. The dual slit experiment proved the collapse of the wave function and the significance of the mind. And, Buddhism is the study of the self and reality. These are related concepts and synthesize perfectly in our playing The Glass Bead Game.

"We can speak and think only of what exists. And what exists is uncreated and imperishable for it is whole and unchanging and complete. It was not or nor shall be different since it is now, all at once, one and continuous . . . "

— Parmenides

"Ex nihilo nihil fit" (Nothing comes from nothing)

— Parmenides

Zeno—Paradoxes

The Achilles Paradox

The Achilles Paradox states that you can never catch a moving object since the leading object will have moved ever so slightly forward every time the trailing object reaches the prior location of the leading object.

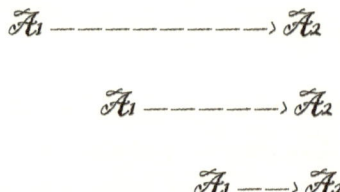

Even though Achilles is extremely fast he will never be able to catch a slow tortoise due to this infinite regress approaching an ever decreasing limit. Through this paradox Zeno proved that reality must be discrete since we must land on only allowable units of reality which in effect jump over all of the little infinities on the way.

The Dichotomy Paradox

The Dichotomy Paradox states that you can never reach a stationary point since you will have to travel one half of any remaining distance prior to reaching your final destination into an infinite regress.

$$A_1 \text{———————} \tfrac{1}{2} \text{————} \tfrac{1}{4} \text{——} A_2$$

No matter how small the remaining distance you will always have to reach the one half-way point prior to reaching your ultimate destination. Therefore, you can never reach your final destination.

Again, this paradox proves that reality must be discrete and not continuous since we must just jump over the infiniteness of the line to reach our final destination.

The Stadium Paradox

The Stadium Paradox states that time cannot be discrete since two space-time points could pass each other in no time at all.

A_1 A_2 A_3

B_1 B_2 B_3 (row travelling.————————right)

C_1 C_2 C_3 (row travelling.————left)

A_1 A_2 A_3

B_1 B_2 B_3

C_1 C_2 C_3

At no time did B_3 pass C_2 if the only allowable perceivable times are discrete units of A_1, A_2, and A_3.

Through this paradox Zeno proved that the world must be continuous and not discrete. This is contrary to his finding in the Achilles and Dichotomy Paradoxes in which he proved that the world was discrete.

The Arrow Paradox

The Arrow Paradox states that at any one discrete moment in time a moving arrow is motionless.

This paradox proves that reality must be continuous since a space-time unit could never be stationary and discrete if it is actually in motion.

The Place Paradox

The Place Paradox states that the place in which the Universe sits must not be a place since all places require another place to sit into an infinite regress. There would be no foundation for "place" to sit on. Therefore, the Universe must sit in a place that is non-place. It must sit outside of Space!

Megna's Parallel Lines Paradox

Megna's Paradox claims that is impossible for two intersecting lines to become parallel since they will intersect out into infinity.

Consider the following intersecting lines which extend in theory out to infinity in both directions:

Attempt to rotate them to start to get them towards the parallel position such as:

What is the defining point at which they no longer intersect and become parallel such as the following?

The only explanation for Megna's Paradox is that infinity merges with zero at a point where there is no longer any intersection. Something that is logically impossible occurs in the most routine fashion!

Another strange thing about this paradox is that Euclid could not prove the 5th Postulate in the elements which stated that two parallel lines never intersect. This postulate has not been proven to this day but is stated as a fact known to be "intuitively true." It is as Kurt Godel stated that some things we take to be true but cannot be proven by their very nature.

Or, what if it is true that there are no such things as parallel lines since infinity is not provable either. Then Megna's Paradox wouldn't be a paradox anymore since intersecting lines would never disengage and become parallel. They would always stay intersected no matter how far out towards infinity you approached.

Megna's Billiard Balls Paradox

Another similar paradox arises when you consider the deterministic cause and effect billiard ball universe in which we live.

At what exact point does the following moving object on the left collide with the object in the middle and cause it to collide with the object to the right of it.

The paradoxical answer is that there is no point!

It can get infinitely close to the object in the middle but never actually collide with it.

Again, something as common as shooting pool balls into one another is logically impossible according to our laws of calculus which only permits objects to approach a limit of actual intersection and contact. It can get infinitely near but never actually get there.

Again, this must mean that the balls collide at an approximated discrete point in the real world.

Mathematician Joseph Mazur explained it to me once that there is a point on top of the infinite regress that allows it to get to its destination. He claims that this is what explains away all of the Zeno like paradoxes. I personally don't quite understand it myself since you still have to traverse space itself which lies below this "discrete infinity hopping."

Douglas Hofstatder — Godel, Escher, Bach

The book Godel, Escher, Bach takes The Glass Bead Game to an even higher level. It synthesizes mathematics, art, and music into a beautiful "golden braid." It uses analogy, pattern, philosophy, art and the laws of science and applies it to the broader question of an ultimate "theory of meaning." It analogizes between the meaningless concepts such as the letters in the alphabet relationship with words which we assign meaning to. This is similar to how our base constituents such as atoms, molecules and so forth which in themselves are meaningless all form to create the self. It is something simple which organizes it in such a way as to become something meaningful.

Aristotle—Physics—Realism—Metaphysics

Aristotle believed in a finite realism in which the human mind can have an ultimate understanding of which he laid out in "Physics." Reality is discrete. He claimed that infinity does not exist since it is impossible and creates all of the paradoxes of logic. Current physicists use a technique called renormalization to get rid of any infinities that mess up the equations to explain reality. All of our logical systems must assume that the universe is limited and discrete so that we can understand it.

Ayn Rand stated that Aristotle was the only philosopher that she admired and was influenced by. He believed that A equals A period. Ultimate Reality was conceivable through the human mind. There were no contradictions or illusions if man believed in the power of his reason. Concepts are accepted as true by the language and definitions that we give them. Realism is the world in which we live. Using this form of logic and reasoning you avoid ever getting caught in any circular arguments which cannot be solved. Questions like is there a God, therefore, aren't even relevant to a realist since it is an unanswerable question.

Anaxagoras—Everything is in Everything

Anaxagoras stated that "everything is in everything" in about 450 BC. This is supported by the idea that a finite set contains infinity and the view of the universe as a holograph in which each piece of reality contains the entirety of the whole within it.

When we think of infinity contained in a finite set it would seem that it equates with an infinity from outside of it as well. How can one infinity be different than another? Georg Cantor claims that he proved that they are different since some infinities are larger than others. As a matter of fact he claimed that some infinities are so large that they make other infinities approach a limit of zero in size. Picture one set (1,2,3,4 . . .) compared with all the fractions in between these in another set. They both are infinite but the latter is so much larger that the first disappears in comparison.

Bertrand Russell— Continuity

All Mathematics can be proven within the system. This he showed in his "Principa Mathematica." Bertrand Russell and Alfred North Whitehead attempted to prove all of mathematics axioms, theorems, and postulates in this book so that there could be no question as to its validity. Kurt Gödel, however, proved with his Incompleteness Theorem that any closed system such as mathematics could not be proven at its most fundamental level. At its foundation it must be proven by something outside of itself. This put mathematics in a quandary for many years since it was thought that it had almost a Platonic existence which had to be valid. David Hilbert stated that if mathematics is not true that man is going to be in trouble because we won't be able to tell if anything is logically true.

Kurt Gödel—Incompleteness Theorem

Kurt Gödel proved in his "Incompleteness Theorem" that all systems including mathematics, set theory, logic, and existence must be proven from outside of the system. The foundation cannot be proven due to an infinite regress of trying to prove something with itself. This verifies the possibility of an existence of some transcendent reason such as a god or some other reason and explanation for a system transcendent of itself. This concept put the foundation of all mathematics on shaky ground since not all of the theorems and axioms could be fully proven to be true.

If we compare the following charts we find a yin-yang scenario as the explanation of being that is counter intuitive according to Kurt Gödel's Proof if the reason and ultimate explanation must come from outside of the original system.

Totality of the Universe	Explanation of the Universe
Being	Non-Being
Body—Matter	Non-Matter
Mind	Non-Mind
Space	Non-Space
Time	Non-Time

The Liars Paradox

Consider the following sentence:

"This Sentence is false"

If the sentence is true it is what it says and it is false.

If the sentence is false it is true that the sentence is false

Consider the following set theory statement

"A set contains all the members which are not in the set."

A circular argument is started since the set cannot contain any members, yet, that is exactly what it states it is.

These are examples of how our logical system breaks down at certain points. This is similar to calculus how it works great up to a point but doesn't quite get us to complete understanding since it only approaches a limit.

The incompleteness theorem can be take even further and applied to existence as a whole such as the following statement:

"I am not provable"

Since Godel's theorem claims that we cannot prove a closed system at a fundamental level from within itself we are left with this question again of "who am I."

It is the craziest thing to think that we can define other constituents within the universe but we can't define ourselves.

The question then arises: How well are we really defining reality?

Werner Heisenberg—The Uncertainty Principle

Knowledge of a particles position and momentum cannot be known at the same time. The more that you know about a particle's position the less you know about its momentum and vice versa. This principle further signifies the elusiveness of reality.

By its very nature it cannot be fully known.

Heisenberg's Uncertainty Principle

Picture the particle below:

O

You cannot have any knowledge of which direction this particle will move since its stationary position is known 100%.

Picture the moving particle below:

————————————————————→

You cannot have any knowledge of this particles exact position if you know 100% the direction and momentum in which it is moving.

This is not a short coming of science but an actual feature of the nature of being.

Erwin Schrodinger—Schrodinger's Cat—Superposition

Erwin Schrodinger proved that a particle or any matter can exist in a state of superposition prior to it being perceived by a human mind. It exists in a dual state of potential. Therefore, he believed that a cat or any being can exist both dead and alive simultaneously prior to being perceived. Being exists in this hybrid like state.

This experiment also raises the question of a "Many Worlds Theory" since you have to ask what happens to all of the other potential photons that don't get quanta sized? Where do they go? Don't they have some sort of existence as well?

If foundational particles exist in this dual state of superposition and potential prior to being perceived does that mean that our being itself possesses this unlimited nature as well? It must!

Georg Cantor—Infinity

Georg Cantor proved that some infinities are larger than others. Also, that some infinities are so large that other infinities approach a limit of zero size and cease to exist.

He argued that there are omega points in which you can quantify an infinite amount of infinities and do mathematical equations substituting the definition of finite quantities with the infinities. This is, again, that idea of a finite set containing an infinite amount of members.

"The study and mere thought of infinity is one of the most fascinating thoughts a man can have. The imagination of it is unlimited!

F. H. Bradley—Appearance and Reality

F. H. Bradley pursued an absolute view of reality.

F. H. Bradley had a significant impact on The Metaphysics of Being which sought an "absolute" non-relative, non-contradictory explanation to existence. It may not be possible in the real world to do this since Einstein proved that we all live in a relative universe. Our Being itself has meaning possibly only as it has relation to the universe and other beings.

I'm not entirely sure that death and non-existence isn't the bottom line finding of some of the idealist views of Buddhism, Platonic Absolutes and The Metaphysics of Being. Unfortunately, as crazy as it sounds it may only be through death and non-being that beings fully merge with the oneness of all infinite reality, eternity, and Truth in which you can say something is absolute and perfect. Certainly the reality which we are perceiving and experiencing is imperfect, contains evil, and is limited to name just a few of its minor flaws.

George Berkeley—To Be is To Be Perceived

George Berkeley believed in an idealist view of reality in which its existence was dependent on a human mind perceiving it. He coined the phrase "to be is to be perceived."

It would seem that Berkeley was proven to be correct that "to be is to be perceived" with the advancement of quantum mechanics. The double slit experiment proves that the mind creates matter with the collapse of the wave function. It creates reality! Therefore, the mind does have a special status in the universe as so many have alluded to. Physicist Max Tegmark believes that we may be the only beings in our galaxy or possibly even the universe to have a mind that can reason and conceive of itself. He goes through an elaborate explanation as to why he believes this is so. Given this special status of the mind alone is reason enough to attempt to preserve our survival as a species. He believes that we may be at that crucial turning point where we either exponentially grow as beings and seek perfection or perish by destroying ourselves.

We, in effect, will either pursue Buddha or Bombs!

Gottfried Leibniz—Monads

Gottfried Leibniz believed that the most fundamental unit of being was a monad which was eternal, had no extension, and was unique. It existed outside the limits and extension of space and time.

This extensionless monad which Leibniz speaks of reminds me of Euclid's— "A point is that which has no extension." The question was how does a line become a line if its foundation is a nothing? Moreover, how does being become something if its foundation is a nothing?

This also reminds me of Descartes Mind—Body Dualism which asks how does something with extension get influenced by something that doesn't have extension?

These are all examples of how The Glass Bead Game is played in which analogies are drawn between various viewpoints. Douglas Hofstadter talked about this type of synthesizing a lot in his book Gödel, Escher, Bach as well. He referred to it as the "Infinite Golden Braid."

Euclid— "A Point Is That Which Has No Extension"

Euclid stated that "a point is that which has no extension." A paradox arises when you consider the existence of a line depends on a series of points which don't exist. Therefore, lines consist of other lines. At the very foundation of mathematics, our logical system, and being itself seems to be this unexplained nothingness as its source and explanation of its existence. This concept of nothing, therefore, has somewhat of a dual meaning of actually meaning nothing and also meaning a something of nothing that has significance in the entirety of the universe.

Euclid's first definition of a point also had a major influence on The Metaphysics of Being. How can we understand the most fundamental reason for "Being" if it doesn't have any attributes? Alan Watts and other Buddhists would say you can't so just enjoy the ride. Ayn Rand and the other realists would also say that there is no sense to even asking this question since it doesn't have any bearing on our immediate perceivable "useable reality."

Democritus—Atomism and Void

Democritus believed in an atomic discrete view of reality coexisting within a void. This is almost a yin-yang scenario in which something and nothing must coexist together. One cannot exist without the other much like light without darkness and good without evil. The atoms themselves, however, are viewed indivisible and eternally indestructible and are considered to be everywhere. They are separated by nothing. The nothingness was necessary according to Democritus so that real matter would have room to move. It was really only the atomic discreteness that was everywhere that was the only concern for man's existence.

Neils Bohr expanded on this concept in 1913 when he defined the atom as having a structure similar to the planetary universe with the sun and its orbits. Today's quantum mechanics describes this more as a cloud of potential not at all similar to the solar system. This is an example of how scientific models are based on our conceived reality of the time and not necessarily related to actual truth.

Albert Einstein—Relativity

Albert Einstein proved that reality is relative to our view point. Nothing exists outside of its relation to something else. Many of his findings were counter intuitive.

He linked space and time which proved the existence of a fourth dimension.

He proved that motion impacts space and time.

The faster something moves the shorter it becomes.

The faster something moves the heavier it becomes.

The faster something moves the slower time is. The Twins Paradox proved that two identical twins travelling at different speeds through the universe could age at significantly different rates.

He theorized that if you could travel at the speed of light you would have no extension, infinite mass, and time would stand still!

$E=MC_2$ showed the equivalence of energy to mass. A small amount of matter contains a massive amount of energy.

The implications of Einstein's thought experiments proved to be one of the greatest discoveries of all time. Even though he was also deeply involved in the idealism of quantum mechanics I think he preferred the sensible deterministic world that Newton had created years earlier. Einstein with the aid of Kurt Gödel tried to explain the "spooky action from a distance" nature to the universe all the way up to his death.

He could never get over the indeterminate nature of reality which consisted of features like wave-particle duality, superposition, entanglement, and so forth. As he stated "God does not play dice."

John Paul Sartre—Being and Nothingness

John Paul Sartre claimed that there were three states of being.

Being-In-Itself was non-conscious being which a nothingness which transcended the conscious mind was.

Being-For-Itself was the conscious mind in which we perceive reality and the nothingness below it.

Being-For-Others was the deterministic and relative world in which we live.

Sartre believed that it was the conscious mind that could conceive of a nothingness below it that gave the real world the foundation to exist. Something exists within a nothingness by its very nature. It is bounded by nothing!

Gary Zukav—The Dancing Wu-Li Masters

Gary Zukav merged the beliefs of eastern mysticism and quantum mechanics. He was not a physicist but was able to tie together both of these complicated concepts in a very easy to understand way. It emphasized the importance of the mind as being the central cause of the universe. The "I" and the "self" which by themselves are not wholly definable are the main tool for our understanding and creation of the universe. Wu Li has many translations from Chinese to English but my favorite one is enlightenment since it sums up in one word this higher state of consciousness which The Glass Bead Game seeks.

John Searle—Philosophy of Mind

John Searle teaches about the philosophy of mind at Berkeley University. He does not believe in Rene Descartes Mind-Body Problem not only since it has ties to mysticism and religion but for more concrete reasons as well. He believes in a more realistic approach to studying the mind and reality. He states that something without extension can't have an influence on something that does. Therefore, he views brain function as definable waves of matter similar to the physiological processes of the body. One of the quotes that I liked the most from John Searle was that his axiom was that if we can conceive of something then it exists.

Brian Green— String Theory and The Elegant Universe

Brian Green believes that the most fundamental unit of being is a one dimensional string which exists in eleven dimensions in a state of perpetual motion. It is the foundation of the cosmic symphony in which we live.

He believes similar to Anaxagoras that "everything is in everything." He asks the question how the smallest of units can be identical to the largest in the universe. He believes in almost an illusionary reality from the one we are currently conceiving and defining. He certainly believes that man is on the verge of a potential for intellectual growth within the coming years. The test going on at Cern may validate a lot of his views which may seem crazy to the general public which is not exposed to the strange quantum mechanical world that he is.

Immanuel Kant—Critique of Pure Reason

Reason—Transcendence

Immanuel Kant did not believe that a limited human mind could understand the entirety of the universe since it transcended it in a noumenal world similar to the Platonic Forms.

Kant's Fore Antimonies

The First Antinomy (of Space and Time)

Thesis: The world has a beginning in time, and is also limited as regards to space.

Kant argued that the "phenomenal" world must be limited, otherwise, it would take an infinite amount of time to reach the time that we are currently living in which is impossible.

Anti-thesis: The world has no beginning, and no limits in space; it is infinite as regards to both time and space.

The Second Antinomy (of Atomism)

Thesis: Every composite substance in the world is made up of simple parts, and nothing anywhere exists save the simple or what is composed of the simple.

Anti-thesis: No composite thing in the world is made up of simple parts, and there nowhere exists in the world anything simple.

The Third Antinomy (of Freedom)

Thesis: Causality in accordance with laws of nature is not the only causality from which the appearances of the world can one and all be derived. To explain these appearances it is necessary to assume that there is also another causality, that of freedom.

Anti-thesis: There is no freedom; everything in the world takes place solely in accordance with laws of nature.

It is interesting that Kant compared a mixture of determinism-freedom in the thesis with the anti-thesis of strict determinism. It seems to me that there should also be a another thesis that cites "pure freedom" which would not be restricted by any determinism of the laws of nature at all and, therefore, be completely free.

The Fourth Antinomy (of God)

Thesis: There belongs to the world, either as its part or as its cause, a being that is absolutely necessary.

Anti-thesis: An absolutely necessary being nowhere exists in the world, nor does it exist outside the world as its cause.

David Hume—Metaphysics

David Hume was an empiricist and skeptic in which he believed that it was human passions and direct experience with nature that explained the universe. In his Treatise on Human Nature he gets into a discussion on whether he believed the world was discrete or continuous. He maintained that it must be discrete for us to make sense of the world. We need to quanta size our conceptions since that is how we form language and ideas so that we can have an simple understanding of a complex universe. This is similar to how Ayn Rand viewed the world. We need to form concepts the are concrete so that we don't get caught in a quagmire of circular arguments where progress is not possible.

Max Planck—Planck Units—Discreteness

Max Planck said that reality is discrete and that it can be perceived only on allowed units that exist within the variable units of h. He believed that there were absolutely smallest units in nature in which nothing else could exist below them. These included Planck units of space and time in which the universe proceeded at the fundamental levels. Below these base levels space and time was meaningless and did not exist. How can there be a place where space and time does not exist? It is a nothingness it would seem that engulfs the something in which we live. I agree with his findings. How could it be any other way in a discrete world? There has to be something which is most fundamental and I think these Planck units are as good of a candidate as anything or at least the idea of a discrete base unit if not the ones he actually describes. Without this bottom level of constituents our logical system drops into an abyss of confusion and inconsistency.

Henri Bergson—Intuition

Henri Bergson believed that our limited discrete minds could not understand the totality of the universe unless we used intuition which was an artistic view of the world that didn't quanta size it.

This view is almost like a hybrid of discreteness and continuity. It is something in between called intuition that can't be fully proven but must be taken as fact for us to have any bearing on an understanding of the world. Some philosophers say that there are paradoxes that are real and that can't be solved no matter how far you get into them. We still need some method for a belief system that gives us a chance at moving forward towards our understanding the ultimate nature of reality.

Charles Siefe—Zero—A Biography of a Dangerous Idea

Charles Siefe wrote Zero—A Biography of a Dangerous Idea which showed the extreme relationship between zero and infinity.

Megna's Paradox shows this relationship. Intersecting lines will continue to intersect out to infinity until which point they seem to merge with zero and become parallel. This proves that the concepts of infinity and zero are equivalent. Zero can be thought of as a negation of something else whereas infinity includes all of the finite members within it. Metaphysics of Being went through an extensive argument of the equivalence between not only zero and infinity but also one. It was at the foundation of the entire book.

He believes that we live within a cosmic computer program much like the movie the Matrix. The interesting thing about quantum computing is that the amount of information that it can potentially generate with only forty entangled pairs computing in states of superposition may exceed that of the entire universe which is ten to the power of one hundred and twenty-seven. Something extremely weird is probably going to happen at that point according to Paul Davies.

Paul Davies—About Time

Paul Davies is a modern physicist who believes in realistic explanations to the universe. He believes that the cause and reason for the universe exists within the perceivable system. He also proposed an idea for an idea in which he believes that the past may not be fully settled yet since it has not been entirely perceived. This is supported by the double slit experiment and the delayed choice experiment.

Imagine viewing light coming from a distant star billions of light years away and all of the choices that light has to travel to get here to earth similar to the double slit experiment. When we view that light he believes that we create a concrete reality that may get sent back in time billions of years to actualize its reality. These thought experiments prove the extreme uniqueness and status that the human mind has. It is almost as if the mind is creating all of reality much like Berkeley talked about in his idealism.

Joseph Mazur—The Motion Paradox

Joseph Mazur revived Zeno's Paradoxes in his book the Motion Paradox. The very fact that he wrote this book as an active Mathematics teacher proves that Calculus has not fully explained this mystery since calculus only approaches a limit.

I asked Joseph Mazur once about Megna's Paradox and how intersecting lines could logically become parallel. He stated that is because we visualize a discrete point on top of the infinity to get out of the continuous logical loop which an infinite regress puts us into. Our finite minds need to have this discreteness to understand things even if the nature of the universe is infinitely continuous in some way below it.

Rene Descartes— "I think, therefore, I am"

Rene Descartes coined the phrase "I think, therefore, I am" since it was the only thing in the world that he knew for certain. He proved his existence every time that he had a thought. His Mind-Body Problem is still debated today. The paradox arises when you consider that matter which has extension can be affected by a mind which has no extension.

Many physicist today would say that the mind has extension just the same as all of matter. Brain waves are no different than the wave-particle duality of all matter. This still doesn't explain thought within these waves in which it can perceive itself outside the physical world in some way.

Physicists are trying to decode these electrical impulses in order to identify what someone is thinking when a certain sequence of synapse firing takes place. In affect they are trying to read people's minds in an attempt to further understand the mind. This sort of advancement of technology exemplifies what Aldous Huxley talked about as being "The Brave New World."

Fritjof Capra—The Tao of Physics

Fritjof Capra's The Tao of Physics merged the study of the Chinese religion of Tao and Western quantum mechanics and sciences. This again was similar to Gary Zukav's The Dancing Wu-Li Masters. Both of these books explain the phenomenon and relationship between the mind and quantum mechanics. The Dao seeks to view the world in a wave like manner since one of its main goals is to become one with the universe through attempts to shed the ego. Many identify the mind with the self and the ego so I'm not sure that this can actually be done in practice. I have talked about magical experiences where the mountain climber becomes one with Mount Everest by losing himself through his focus with the event. These may be instances where the concept of time is lost and the perceiver and the perceived seemed to merge much the same way that quantum mechanics dictates. Other magical examples of this may be love making, creating music or other forms of art, Olympic lifting events, or as Alan Watts would claim just peeling potatoes can attain this nirvana as well.

Julian Barbour—The End of Time

Julian Barbour's The End of Time theorized that the space-time units in which we perceive are actually eternal discrete snap shots that are viewed in a sequence similar to a movie presentation in which the still shots of slides are moved to create the illusion of motion in a film.

This is similar to Leibniz's monads in which these fundamental units are dimensionless and eternal. This is different from Planck's view which believes that they have some minimal extension. Julian Barbours transcendent forms of something live in a world of nothingness which again is beyond human conception. So, the argument continues to go round and round in search of the ultimate "Shiva" as the source of our being. Possibly the work at Cern in Geneva, Switzerland will further our understanding of this. I used the term Shiva since that is the sculpture that is out front of the Cern accelerator. Shiva is the Hindu God which stands for the source of energy of the universe.

Stephen Hawking—A Brief History of Time

Stephen Hawking explained for the general public how black holes transcend the rules of our perceived world in which time and space do not exist. The fact that black holes exist which don't possess the same rules as the world we live in justifies some of the crazy possibilities of the future of the mind and of man. This next "transhuman" evolution that is going to take place is coming on similar to an exponential curve. We are proceeding along at an upward pace as a civilization but at some point this is going to take off into some sort of "The Matrix" type world where the mind will merge with computers, stem cells will cure all diseases, quantum computers will exceed the capability of the entire universe and we will attain that eternal utopia that everyone has been seeking for thousands of years.

John Mctaggart—The Unreality of Time

John Mctaggart proved mathematically and logically that time cannot exist. He believed that Metaphysics was a higher discipline than science and physics since it pursued the ultimate explanation of reality.

F. H. Bradley was an important influence on McTaggart. McTaggart thought that Bradley was "the greatest of living philosophers" and once told G. E. Moore that when Bradley walked in, "he felt as if a Platonic Idea had entered the room. His arguments were similar to Zeno's in that he proved that our logical system itself is what verifies that time cannot exist. It breaks down at its foundation contrary to the world that we think we are perceiving.

Roger Penrose—The Emperors New Mind

Roger Penrose wrote the Emperors New Mind in which he describes a higher state of consciousness through the new understanding of science and reality. He is one the most respected physicist today and also verifies that we are approaching a new age. It is interesting that he stresses that we need to keep in my mind the other branches of philosophy which have concerns for beauty and morality as well. We can't only be concerned about the advancement of technology or we may find that it ends up destroying us instead of helping us.

Alfred North Whitehead—Process and Reality

Alfred North Whitehead explained the process in which we perceive reality similar to the eastern ways of thinking. He believed in a timeless entity as the foundation of all reality. He helped Bertrand Russell with the writing of Principa Mathematica in which they tried to prove all of the axioms, postulates, and theorems of all of mathematics to verify its validity. At the end of the day it seems it is as if some other transcendent rule must be needed to explain the foundation of everything. Again, a closed system was not able to prove itself at its very foundation. This is one of the most puzzling things when you consider similar to the self trying to conceive and understand the self. It just goes round and round and never can quite get to the bottom of it.

Richard Bach—Jonathon Livingston Seagull—Limits

Richard Bach's Jonathon Livingston Seagull and Illusion's described how the limits of existence can be transcended. This is a simple story of trying to exceed limits through the use of one's wings. In final analysis Jonathon Livingston Seagull realizes that he has attained Nirvana by the mere fact that he has the joy of flight and that the final destination is really of no concern.

Isaac Newton—Laws of Motion

Isaac Newton's classical physics is the deterministic reality in which we live our day to day lives in. It is a billiard ball existence of cause and effect and common sense. He proved how you can take a simple thing like an apple falling from a tree and apply its universal principles to the entire universe. Our minds are able to perceive patterns which have a certain beauty and elegance to them that help us understand the world in which we live. The artists, philosophers, and scientists of today, however, are finding new realities which are even better and more powerful than those identified in the deterministic rules of Newton. It is the creative indeterminate world which is being unharnessed that is leading us into the new age. This will be a future world in which we can be truly free!

Michelangelo—The Pieta, The Sistine Chapel, and The Statue of David

Michelangelo was one of the greatest painters and sculptors who ever lived. The beauty of his works have been admired by millions of people ever since. The creativity he presented revealed the beauty that existed in an eternal sense. He claimed that the Statue of David was already there within the marble and he just exposed it for everyone to see. This eternal beauty can be applied to the universe in which we live as well. Currently we see imperfections in the universe which contains evil, suffering, disease, wars, and so forth. What if there is a world that is perfect that exists but we are for some reason not realizing it yet!

John Keats—Beauty is Truth, Truth Beauty

John Keats wrote the Ode on a Grecian Urn in which he stated that "Beauty is Truth, Truth Beauty—that is all ye know on earth and all ye need to know."

Love is my religion—I could die for it.

John Keats

My imagination is a monastery and I am its monk.

John Keats

Nothing ever becomes real till it is experienced.

John Keats

Lisa Randall—Warped Passages—Hidden Dimensions

Lisa Randall believes that many dimensions exist outside of our current perception similar to the beliefs of Brian Greene and his views on string theory. She recently wrote a book on perspective in which she explains that it is our artistic relative perspective view of reality that limits our thinking and suggests the impossibility of many of the findings of modern science and quantum mechanics. It is this creative artistic and mathematical view of the world which affords us a greater understanding of the universe.

It's hubris to think that the way we see things is everything there is.

LISA RANDALL, Discover Magazine, July 2006

Although I was first drawn to math and science by the certainty they promised, today I find the unanswered questions and the unexpected connections at least as attractive.

LISA RANDALL, Warped Passages

Thomas Nagel—A View from Nowhere.

Thomas Nagel wrote A View from Nowhere in which he explains that our view of the universe is limited since we cannot see the totality of existence all at once. We do not possess the capability of a "gods view" of the universe in which we can see the entire circumference of a three dimensional object all at once. We see only one side of reality at a time.

Consciousness is what makes the mind-body problem really intractable.

Thomas Nagel

Life may be not only meaningless but absurd.

Thomas Nagel

There is a tendency to seek an objective account of everything before admitting its reality. —Thomas Nagel

Thomas Young—Double Slit Experiment

Thomas Young's Double Slit Experiment in 1805 was the foundation for the understanding of quantum mechanics more than 100 years later. This proves the wave-particle duality, superposition, and, that the mind creates reality through the collapse of the wave function. It may even hint at a "many worlds theory" since we have to wonder what happens to all those particles of potential that never become "real."

——————→   ——————→   (Actual Existence)

                          ——————→   (Potential Existence)

We know that potential existence is real since a wave pattern is formed while particles are in a state of superposition prior to being perceived! Therefore, "The Many Worlds Theory" has to be real. If our base constituents have this characteristic our entire existence must have this as well.

John Stuart Bell—Incompleteness Theorem—Entanglement

John Stuart Bell's Incompleteness Theorem proved the holistic nature of the universe. Entangled pairs of particles allow for the knowledge of the distant pair no matter how far away it is in the universe. This knowledge in a sense exceeds the speed limit of the universe, light, even though no information is actually physically transmitted between the two particles.

$$\circ \quad \ll\text{——————————————————————}\rightarrow \quad \circ$$

"Spooky action at a distance"

Louis DE Broglie—Wave-Particle Duality of Matter

In 1924 Louis DE Broglie proved that all subatomic particles possess a wave-particle duality. Particles exist in a state of potential as waves prior to being perceived at which time they are quanta sized into an actual discrete particle. Some people believe that all matter including human beings may possess this dual nature to their existence but for some reason it is not perceivable to the human mind.

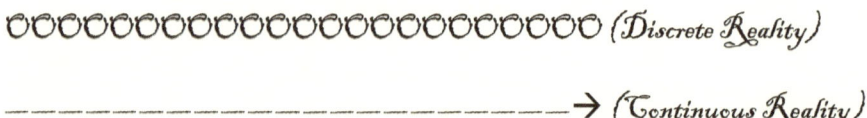

OOOOOOOOOOOOOOOOOOOOOOOOOOOO (Discrete Reality)

————————————————————————————→ (Continuous Reality)

The continuous reality is really discrete reality as well if you look close enough. True continuity does not exist in the world which we are currently perceiving. It only exists in some sort of theoretical sense at some transcendent level below our conceived limited understanding.

John Archibald Wheeler—Law without Law and Black Holes

John Wheeler coined the phrase "law without law" in which he believed that at the foundation of all existence may be an entirely unrecognizable set of laws contrary and counter intuitive to ways we are currently viewing the world. This is an interesting concept since quantum mechanics already is counter intuitive to the classical mechanics of Isaac Newton and the everyday world in which we live.

John Wheeler also developed the thought experiment called the "delayed choice experiment" in which he expanded on the ideas of the double slit experiment to included viewing a particle after it had already chosen which slit to pass through. It adds another choice to the experiment after the particle is collapsed and quanta sized and, therefore, allows it to "go back in time" and reemerge in a state of wave like potential. This experiment also verified that extreme uniqueness of the human mind and its ability to create reality not only by actually perceiving but also by an apparent mere "ability to know" as well while the particle remains unperceived. Behind it all is surely an idea so simple, so beautiful, that when we grasp it—in a decade, a century, or a millennium—we will all say to each other, how could it have been otherwise? How could we have been so stupid?

John Archibald Wheeler

If you haven't found something strange during the day, it hasn't been much of a day. John Archibald Wheeler

In any field, find the strangest thing and then explore it. John Archibald Wheeler

In order to more fully understand this reality, we must take into account other dimensions of a broader reality.

John Archibald Wheeler

Neils Bohr—Theory of the Atom

Neils Bohr created the theory of the atom which is currently being taught in our schools. This verified the atomic belief that Democritus had presented in 400 BC that the world is discrete existing within an external void.

Everything we call real is made of things that cannot be regarded as real.

Niels Bohr

If quantum mechanics hasn't profoundly shocked you, you haven't understood it yet. —Niels Bohr

Technology has advanced more in the last thirty years than in the previous two thousand. The exponential increase in advancement will only continue.

Niels Bohr

Bryce DeWitt—Many Worlds Theory

Bryce Dewitt proposed the many worlds theory in the 1940's prior to the development of string theory and other theoretical advances in modern physics.

"No development of modern science has had a more profound impact on human thinking than the advent of quantum theory. Wrenched out of centuries-old thought patterns, physicists of a generation ago found themselves compelled to embrace a new metaphysics. The distress which this reorientation caused continues to the present day. Basically physicists have suffered a severe loss; their hold on reality."

Bryce DeWitt

Hugh Everett—Many Worlds Theory

Hugh Everett also proposed the many worlds theory as his theory of everything but left the field of physics since no one took him seriously at the time so he left his vocation for the business world.

Hugh Everett's work has been described by many people in terms of many worlds, the idea being that every one of the various alternative histories, branching histories, is assigned some sort of reality. —Murray Gell-Mann

Leo Buscaglia—Love

Leo Buscaglia was an education professor at USC who developed an extensive theory of love in which he gained a world following for. He believed that our university system was failing for the primary fact that we teach many things to students but not the most important thing in the world which is love.

"A world lived for tomorrow will always be one day away from being realized."

Leo Buscaglia

Death is a challenge. It tells us not to waste time . . . It tells us to tell each other right now that we love each other. —Leo Buscaglia

I have a very strong feeling that the opposite of love is not hate—it's apathy.

It's not giving a damn. Leo Buscaglia

Love is life. And if you miss love, you miss life.

Leo Buscaglia

Erich Fromm—The Art of Loving

Erich Fromm wrote the Art of Loving which proposed the idea that loving is an art and discipline which must be learned and practiced similar to learning the art of medicine. He cited four main features of active love which were care, responsibility, knowledge, and respect.

If I am what I have and if I lose what I have who then am I?

Erich Fromm

In love the paradox occurs that two beings become one and yet remain two.

Erich Fromm

Love is the only sane and satisfactory answer to the problem of human existence.

Erich Fromm

Heraclitus—Flux—Eternal Change

Heraclitus believed in flux which was that the universe was in perpetual state of change. It was change itself which was the foundation of the elusive nature of reality and being. It is Change that is absolute, eternal, and unchanging.

Change alone is unchanging.

Heraclitus

No man ever steps in the same river twice, for it's not the same river and he's not the same man.

Heraclitus

Georg Hegel—Idealism

George Hegel believed in an idealistic view of the universe in which the mind exists outside the deterministic limits of the universe.

Nothing great in the world has ever been accomplished without passion.

Georg Wilhelm Friedrich Hegel

Truth in philosophy means that concept and external reality correspond.

Georg Wilhelm Friedrich Hegel

Fredrick Nietzsche—Nihilism—Belief in nothing absolute—"Overman"

Fredrick Nietzsche's nihilistic view of the universe led him to state that "God is dead." He developed an elaborate argument against the powers of the state and religion in which he believed that they had propagated a system to control the masses through a lie in which only the state possessed the divine knowledge of the universe. He completely rejected these lies, limits, and controls and sought the development of the "overman" or superman which transcended these dogmatic laws.

And if you gaze for long into an abyss, the abyss gazes also into you.

Friedrich Nietzsche

Jacob Bernoulli—Infinity—A finite set contains an infinite amount of members

Jacob Bernoulli proved that a finite set contains an infinite amount of members. This is at the foundation of many of the paradoxes that exist in our current view of reality.

It is utterly implausible that a mathematical formula should make the future known to us, and those who think it can would once have believed in witchcraft.

Jacob Bernoulli

M. C. Escher—Escher Tiling's—An enclosed image that contains an infinite amount of images

M. C. Escher was an artist who developed the concept of an Escher Tiling which is in theory a finite image which contains an infinite amount of images.

He who wonders discovers that this in itself is wonder. —M. C. Escher

William Blake—Eternity in an hour

"To see a world in a grain of sand,

And a heaven in a wild flower,

Hold infinity in the palm of your hand,

And eternity in an hour."

William Blake

Blaise Pascal—Man emerges from nothingness

Blaise Pascal stated that "man was unaware of the nothingness from which he emerges."

If man made himself the first object of study, he would see how incapable he is of going further. How can a part know the whole?

Blaise Pascal

Martin Heidegger—Being and Time

Martin Heidegger developed a systematic format in which to study the ontology of being itself. He believed that the prior philosophers throughout the history of time stated "being" as a given. In his work Being and Time he proved the existence of the self through Dasein.

The possible ranks higher than the actual.

Martin Heidegger

Why are there beings at all, instead of Nothing?

Martin Heidegger

Aldous Huxley—A Brave New World

Aldous Huxley believed in a higher state of being through the eastern ways of thinking in which we perceive reality as one interconnected system. It will be this new brave world of existence which will take us to the next evolution of man.

Maybe this world is another planet's hell.

Aldous Huxley

Facts do not cease to exist because they are ignored.

Aldous Huxley

Mahatma Gandhi—Non-Violence

Mahatma Gandhi's theory of non-violence made him one of the greatest human beings who ever lived. He was shot and killed for his beliefs in peace.

A man is but the product of his thoughts what he thinks, he becomes.

Mahatma Gandhi

My religion is based on truth and non-violence. Truth is my God. Non-violence is the means of realizing Him.

Mahatma Gandhi

When I admire the wonders of a sunset or the beauty of the moon, my soul expands in the worship of the creator.

Mahatma Gandhi

Jesus—Salvation

Jesus was one of the most important people to ever walk the earth. He has had more of an impact on society than anyone who may have ever lived. Although he preached that we are all sons of God and to be loving towards all men society has corrupted his name to create an elaborate system to control the masses, coerce altruism, and commit many crimes and acts of war.

Buddha—Zen and Nothingness

Buddha, Siddhartha, believed in the pursuit of Nirvana by losing the self while studying reality and the self. He believed that you return to the self after stripping away all limited views of the world and recognizing your oneness with it.

Do not dwell in the past; do not dream of the future, concentrate the mind on the present moment.

Buddha

It is better to conquer yourself than to win a thousand battles. Then the victory is yours. It cannot be taken from you, not by angels or by demons, heaven or hell.

Buddha

David Hilbert— Space

David Hilbert developed the theory of Hilbert Space in which the possibility of many dimensions may exist outside the perception of the mind. Lisa Randall gave credence to his ideas which he developed in the early 1900's.

No other question has ever moved so profoundly the spirit of man; no other idea has so fruitfully stimulated his intellect; yet no other concept stands in greater need of clarification than that of the infinite.

David Hilbert

Johannes Sebastian Bach—Symphony of the Universe

Classical music was studied by Joseph Knect in The Glass Bead Game and was correlated with the precision and beauty of mathematics. You could apply this even further to string theory in which one dimensional strings in perpetual motion may be the foundation of the grand symphony in which we live and call reality.

The aim and final end of all music should be none other than the glory of God and the refreshment of the soul.

Johannes Sebastian Bach

Ludwig van Beethoven—Music Without Sound

Beethoven was one of the greatest composers who ever lived despite the fact that he lost his hearing later in life.

Music is a higher revelation than all wisdom and philosophy.

—Ludwig van Beethoven

Wolfgang Amadeus Mozart—Perfection

I thank my God for graciously granting me the opportunity of learning that death is the key which unlocks the door to our true happiness.

Wolfgang Amadeus Mozart

Jacqueline Du Pre—Cellist

Jacqueline Du Pre was known as the best cello musician in the world in the 1960's but was stricken with M. S. at the age of 28 and, subsequently lost feeling in her fingers.

"Playing lifts you out of yourself into a delirious place."

Jacqueline du Pre

Andre Gregory—My Dinner with Andre

Andre Gregory talked about secret sanctuaries and pockets of light where people could go to escape the over bearing existence of every day society. There they would be able to pursue intellectual and artistic adventures to verify that they are still fully alive.

"A baby holds your hands, and then suddenly, there's this huge man lifting you off the ground, and then he's gone. Where's that son?

Andre Gregory

Do you know, in Sanskrit the root of the verb "to be" is the same as "to grow" or "to make grow".-Andre Gregory

They've built their own prison, so they exist a state of schizophrenia. They're both guards and prisoners and as a result they no longer have, having been lobotomized, the capacity to leave the prison they've made, or to even see it as a prison. Andre Gregory

Stevie Ray Vaughn—Musician

Stevie Ray Vaughn was one of the greatest guitarists of all time.

"What I am trying to get across to you; is please take of yourselves and those that you love; because that is what we are here for, that's all we got, and that is all we can take with us. Are you with me?"

Stevie Ray Vaughan

Glenn Beck—The Gandhi of Our Time

Glenn Beck will go down in history as almost a Gandhi like figure for his role in exposing the underlying unspoken conspiracy against the rights and freedoms of mankind.

And I've come to the place where I believe that there's no way to solve these problems, these issues—there's nothing that we can do that will solve the problems that we have and keep the peace, unless we solve it through God, unless we solve it in being our highest self. And that's a pretty tall order. —Glenn Beck

Neil Young—Cortez The Killer

Neil Young's song "Cortez the Killer" talks about a perfect society of the Aztec's where "all the woman were beautiful."

Better to burn out than rust out.

Neil Young

I just do what I do. I like to make music.

Neil Young

Patti Smith—Rock and Roll Nigger and Poet

Patti Smith started her artistic career as a poet and ultimately became famous at the outset of the New York punk scene of the 1970's. She wrote a song called "Rock and Roll Nigger" in which she described her experience as an outcast of society for not conforming to traditional values. In an interview later in life she stated that The Glass Bead Game was her favorite novel of all time and that she read it about five times so that she could study it in depth and apply its principles to her poetry, art and music.

An artist is somebody who enters into competition with God.

Patti Smith

Orson Wells—Fabian Society

Orson Wells made many predictions about the future of society back in the 1800's which later became true. He was one the first Fabien Progressive Socialist that envisioned a utopian society.

I don't pray because I don't want to bore God.

Orson Welles

Michio Kaku—Buddhism and Quantum Mechanics

Michio Kaku was raised as a Buddhist as a child. Currently he is a theoretical physicist who has made many predictions of the future of mankind. Many of his predictions may seem as science fiction to the general public that is unaware of many of advances in science, mathematics, theoretical physics, quantum mechanics, quantum computing, and so forth.

"Some people seek meaning in life through personal gain, through personal relationship, or through personal experiences. However, it seems to me that being blessed with the intellect to divine the ultimate secrets of nature gives meaning enough to life." Michio Kaku,

George Orwell—1984

George Orwell's 1984 described in detail the insidious nature of the fascist state in which we live. The Orwellian nightmare was predicted as the inevitable state of affairs of a controlling elite class which wants to manipulate the masses through a web of indoctrination, limits, taxes, regulations, norms, and pre-ordained liberal progressive thinking in an attempt to create a utopian society in the image that they have envisioned.

Who controls the past controls the future. Who controls the present controls the past. — George Orwell

Dali—Surrealism

Salvidor Dali was a surrealist painter who attempted to alter our view of reality.

He lived his later years in relative seclusion in order to seek a higher state of consciousness through the use of mathematics. In effect, he created his own Glass Bead Game.

Have no fear of perfection—you'll never reach it.

Salvador Dali

Surrealism is destructive, but it destroys only what it considers to be shackles limiting our vision. — Salvador Dali

Leonardo Da Vinci—Magus and Grand Master of a Secret Esoteric Society

Leonardo was a Magus and Grand Master of a Secret Esoteric Society and his art is filled with codes and allusions to the Sacred Sciences, if you know how to look.

Beyond a doubt truth bears the same relation to falsehood as light to darkness.

Leonardo da Vinci

Norman Vincent Peale—The power of positive thinking

Change your thoughts and you change your world.

Norman Vincent Peale

Imagination is the true magic carpet.

Norman Vincent Peale

Conclusion

This book is a compilation of the greatest minds and ideas of all time incorporated into one synthesis known as The Glass Bead Game located in The House on the Hill. It is a living I Ching in which to inspire future Magister Ludi (Masters of The Game) to create future artistic and intellectual thoughts of perfection. It not only seeks an answer to the biggest philosophical question of all "why is there anything at all" but more importantly recognizes that "Beauty is Truth, Truth Beauty!"

The House on the Hill is an illusionary place were potential profound ideas become actualized and realized. It is the unification of philosophy, art, religion, quantum mechanics, science, music, literature, mathematics, and the mind merged into one collective system known as The Glass Bead Game. It is my Metaphysics of Being.

What does it mean to be truly free?"

Mark Megna

References

The Glass Bead Game By Hermann Hesse Copyright 1943—Fretz and Wasmuth—English translation copyright 1969 Holt, Rinehart, and Winston

Hermann Hesse—Autobiographical Writings Copyright 1971 Farrar, Straus, and Giroux, Inc.

Discourse on Method and Meditations on First Philosophy by Rene Descartes, David Weissman Editor, Copyright 1996 by Yale University

Aristotle—Metaphysics translated by Richard Hope, Copyright 1952 Columbia University, University Press New York

Plato—The Collected Dialogues—Edited by Edith Hamilton and Huntington Cairns. Princeton University Press. 1989.

The Basic Works of Aristotle—Edited by Richard McKeon. The Modern Library. 2001.

Understanding Einstein's Theories of Relativity. Mans perspective on the cosmos.—Stan Gibilisco. Tab Books Inc. 1983. Dover Publications, Inc. 1991.

The Elegant Universe—Brian Greene. Vintage Books. 2003.

Being and Nothingness. Jean-Paul Sartre. Washington Square Press. 1984. Philosophical Library Inc.

Basic Writings of Kant. Edited by Allen W. Wood. Modern Library. 2001.

The Treatise of Human Nature. David Hume. Barnes and Noble, Inc. 2005.

Discourse on Metaphysics. Correspondence with Arnauld. Monadology. G. W. Leibniz. Open Court Publishing Company. 1902.

The Creative Mind. An Introduction to Metaphysics. Henri Bergson. Dover Publications. 2007.

Zero. The Biography of a Dangerous Idea. Charles Seife. Penguin Group. 2000.

Metaphysics of Being by Mark Megna—Xlibris Corporation Copyright° 2009. I.S.B.N: Softcover 978-1-4415-5778-0

Our Knowledge of the External World. Bertrand Russell. Barnes and Noble Inc. 2008.

Three Dialogues between Hylas and Philonous. George Berkeley. Longman. 2005.

The Motion Paradox. Joseph Mazur. Penguin Group. 2007

About Time.—Paul Davies. Orion Productions. 1995. Simon and Schuster.

Meditations on First Philosophy.—Rene Descartes. Hackett Publishing Company. 1979.

Quantum Reality.—Nick Herbert. Anchor Books. 1985.

Jonathon Livingston Seagull.—Richard Bach. The Macmillan Company. 1970.

The End of Time. Julian Barbour. Oxford University Press. 1999.

A Brief History of Time. Stephen Hawking. Bantam Books. 1988.

The Enneads. Plotinus. Penguin Group. 1991.

Zeno's Paradoxes. Edited by Wesley C. Salmon. Hackett Publishing Company, Inc. 2001.

The Infinite Book. — A short guide to the boundless, timeless, and endless. — John D. Barrow. Vintage Books. 2005.

The Emperor's New Mind. — Roger Penrose. Oxford University Press. 1989.

The Tao of Physics. — Fritjof Capra. Shambhala Publications, Inc. 1991.

The Power of Now. — Eckhart Tolle. New World Library. 1999.

In Search of Schrodinger's Cat. — John Gribbin. Bantam Books. 1984.

Warped Passages. — Lisa Randall. Harper Perennial. 2005.

Metaphysics. — Tim Crane and Katalin Farkas. Oxford University Press. 2006.

The Dancing Wu Li Masters. — Gary Zukav. William Morrow and Company. 1979.

Process and Reality. — Alfred North Whitehead. The Free Press. New York. 1978

The Elements. — Euclid. Dover Publications. 1956.

The Nature of Existence. — John Mctaggart. Kessinger Publishing Company. 2008.

Appearance and Reality. — F. H. Bradley. Oxford University Press. 1963.

Loving Each Other. Leo Buscaglia. Random House Publishing Group. 1985.

The Void. Frank Close. Oxford University Press. 2007.

Quantum Questions. Ken Wilber. Shambhala Publications, Inc. 2001.

Cover Design—3D work titled Glass Bead Game Revisited this file was originally published by Cognitive Distortion

www.ingramcontent.com/pod-product-compliance
Lightning Source LLC
Chambersburg PA
CBHW022106170526
45157CB00004B/1505